Internet of Things (IoT) and Wastewater Reuse

Published 2024 by River Publishers
River Publishers
Alsbjergvej 10, 9260 Gistrup, Denmark
www.riverpublishers.com

Distributed exclusively by Routledge
605 Third Avenue, New York, NY 10017, USA
4 Park Square, Milton Park, Abingdon, Oxon OX14 4RN

Internet of Things (IoT) and Wastewater Reuse/by Ana Cristina Faria Ribeiro, A. K. Haghi.

© 2024 River Publishers. All rights reserved. No part of this publication may be reproduced, stored in a retrieval systems, or transmitted in any form or by any means, mechanical, photocopying, recording or otherwise, without prior written permission of the publishers.

Routledge is an imprint of the Taylor & Francis Group, an informa business

ISBN 978-87-7004-201-7 (paperback)

ISBN 978-87-7004-220-8 (online)

ISBN 978-8-770-04213-0 (ebook master)

A Publication in the River Publishers Series in Rapids

While every effort is made to provide dependable information, the publisher, authors, and editors cannot be held responsible for any errors or omissions.

Internet of Things (IoT) and Wastewater Reuse

Ana Cristina Faria Ribeiro

University of Coimbra, Portugal

A. K. Haghi

Chemistry Center, University of Coimbra, Portugal

NEW YORK AND LONDON

Contents

Preface	vii
Abstract	ix
About the Authors	xi
List of Abbreviations	xiii
1 Introduction	**1**
2 Methods	**9**
2.1 Fomulation	11
2.2 Research Approach	12
3 Results and Discussion	**15**
3.1 Regression and Wastewater Reuse	19
3.2 Regression Variables; PHA and Reference Values	19
3.3 Regression Analysis	20
3.4 Wastewater Reuse System Control	28

Contents

4 Conclusion   31

References   33

Index   45

Preface

Today, there is a need for optimal and unified management of water resources in the European continent, and its intensification, especially in the western and southwestern regions. The European Commission with the participation of the European Center pays attention to Environmental Protection, with numerous decrees and directives in the direction of fighting and managing water pollution, such as consumption management and wastewater recycling. Solar desalination systems, virtual water management, and the Internet of Things (IoT) have been developed and approved. This activity is about the management of water consumption for general purposes, prepared by Stab European Union countries, in the field of management and reduction of water consumption based on reducing the use of natural resources and preserving them for the future, and safe living when performing valuable activities and achieving countless successes.

This work is aimed at an IoT model that can be the solution to the problems of water facilities. The paper presents the IoT as one of the efficient methods for the control of a set of water and wastewater networks. Another novelty of this work is that it informs engineers about the state of the use of remote sensing (RS) facilities equipped with networked sensors, advanced modems, data loggers, and the IoT. This can lead to reducing water loss and saving drinking water. The present work also investigated IoT as a high-precision and quick method of incorporation with the RS for wastewater reuse which can lead to polyhydroxyalkanoate (PHA) production and its copolymers as biodegradable plastic. This work finally shows that IoT and RS can be linked to new techniques including a geographic information system (GIS) that could be serious subjects for future research in the fields of environmental engineering, mechanical engineering, electrical engineering, and control engineering in universities and industries.

Abstract

In the 21st century, the boundaries of water and wastewater systems management have gone beyond the technical, engineering, and economic fields. Culture is also an extension of the methods of overcoming the difficulties of moving forward. The water and wastewater systems managers and experts in areas affecting this industry, such as water management. In recent years, the World Bank and other international institutions have focused on the benefits of a water-saving approach. They have provided water savings aggregation and customization of services. By paying attention to the Internet of Things (IoT) a scientific mechanism has led to beneficial results.

Water is an important theme in countries that suffer from water shortages, water stress, and excessive water consumption. Due to its special nature, the water and sewage system in the 21st century is based on the performance of wastewater reuse. This work shows that in the wastewater reuse process, there is a significant relation between polyhydroxyalkanoates (PHAs) and the reference value. Wastewater reuse can lead to degradable plastics that have been prepared by a group called polyhydroxyalkanoates (PHAs) and its copolymers as biodegradable plastic. By regression analysis and through investigation of the scatter diagram and the curve fit due to regression analysis, PHA against the reference value equation showed a good correlation between the dependent variable (PHA) and the independent variable (reference value) data.

Keywords: Internet of Things, anaerobic, wastewater reuse, aerobic, networked sensors.

About the Authors

Ana Cristina Faria Ribeiro, PhD, is a researcher in the Department of Chemistry at the University of Coimbra, Portugal. Her area of scientific activity is physical chemistry and electrochemistry. Her main areas of research interest are the transport properties of ionic and non-ionic components in aqueous solutions. Dr Ribeiro has supervised master's degree theses as well as some PhD theses and has been a thesis jury member. She has been a referee for various journals as well as an expert evaluator of some of the research programs funded by the Romanian government through the National Council for Scientific Research. She has been a member of the organizing committee of scientific conferences, and she is an editorial member of several journals. She is a member of the Research Chemistry Centre, Coimbra, Portugal.

A. K. Haghi, PhD, is a retired professor and has written, co-written, edited, or co-edited more than 1000 publications, including books, book chapters, and papers in refereed journals with over 3800 citations and an h-index of 32, according to the Google Scholar database. He is currently a research associate at the University of Coimbra, Portugal. Professor Haghi has received several grants, consulted for several major corporations, and is a frequent speaker to national and international audiences. He is the Founder and former Editor-in-Chief of the International Journal of Chemoinformatics and Chemical Engineering and Polymers Research Journal. Professor Haghi has acted as an editorial board member for many international journals. He has served as a member of the Canadian Research and Development Center of Sciences & Cultures. He has supervised several PhD and MSc theses at the University of Guilan (UG) and co-supervised international doctoral projects. Professor Haghi holds a BSc in urban and environmental engineering from the University of North Carolina (USA) and holds two MSc degrees, one in mechanical engineering from North Carolina State University (USA) and another one in applied mechanics, acoustics, and materials from the Université de Technologie de Compiégne (France). He was awarded a PhD in engineering sciences at Université de Franche-Comté (France). He is a regular reviewer of leading international journals.

List of Abbreviations

λ_0	Unit of length
γ	Specific weight (N/m^3)
ν	Fluid dynamic viscosity (kg/m s)
A	Constant (0.35–0.55); average ≈ 0.5
B	Constant (0.07–0.15); average ≈ 0.1
B_V	Volume rate (kg BOD5/m^3 day)
B_V	Tank volume rate (kg BOD5/m^3 day)
C	Slope (deg.)
C_S	Amount of water dissolved oxygen in saturation condition (mg/L)
C_X	Dissolved oxygen density (mg/L)
D	Diameter of each pipe (m)
F	Atmospheric pressure coefficient = 1 (at sea level) and for every 1000 (m) height the "F" value decreases by 0.1)
f	Friction
F/M	Coefficient is the food to micro-organism ratio
H	Head (m)
Ha.t	Aeration tank depth (m)
h_L	Combined head loss (m)
h_p	Head gain from a pump (m)
LO	Total influent pollution (kg – BOD5)
MLSS	Mixed liquor suspended solids (kg/m^3)
MLVSS	Mixed liquor volatile suspended solids (kg BOD5/day)
ON	Need of oxygen for nitrogen zings
p	Pressure (N/m^2)
Q	Influent flow rate (m^3/h)
Q_C	Oxygen required per 0.1 (m^3) of aeration tank per 24 hours (kg O$_2$/m^3 day)

List of Abbreviations

Q_{Max}	Max influent in without rain case
Q_V	Amount of oxygen per 24 hours for one cubic meter of aeration lagoon (kg/ m^3 day).
R	Pipe radius (m)
RS	Return sludge
Rt	Detention time (h)
S	Length (m)
SLS	Effluent sludge due to suspended solids (kg)
SMLSS	Sludge mixed liquor suspended solids
SSE	Effluent suspended solids density (kg/m^3)
t	Time (s),
V	Velocity (m/s)
Va.t	Aeration tank volume (m^3)
V_S	Sedimentation velocity (m/s)
y	Aeration tank efficiency

CHAPTER 1

Introduction

Today the increase in the population and consumption of water is one of the biggest challenges of the present world, which could become the cause of many problems in the future. The positive and negative trends of the world have been discussed in the context of water problems in a large scale. Now it stands out on national, regional, and even global scales based on the report of the European Commission, European Union. In the past three decades, it has been increasingly affected by frequent and severe droughts, affected by many environmental factors, especially global changes, between 1991 and 2019. The number of the areas and population affected by drought has increased to approximately 21%, especially in the south-western regions, and the Mediterranean Sea is experiencing one of the most severe droughts, which has affected more than 142 million people, one-third of the total population. It has affected Europe, as a result of the consequences of drought, by a 8.7 billion total loss ($11.4 billion). According to the report of the European Commission, approximately 11% of the population of continental Europe experience water shortages throughout the year, and 23% also suffer from severe water shortages in the hot months.

The extraction of water resources is likely to cause many disputes and conflicts between countries that have common water basins. If there are such changes in this direction, without economic and environmental considerations, the water stress in continental Europe will increase by a significant amount according to the intensification of the water shortage [1–15].

Accordingly, this means the maximum recovery of existing water resources and maximum withdrawal of underground water resources. The European Water

Introduction

Commission, considering the scope of the subject matter, for example, water consumption management in the countries of Germany, France, and the United States are briefly provided with the required data from electronic information systems. These data were collected by the following:

- European Union Biosafety Agency.
- German Federal Ministry of Health and Biosafety.
- German Association for Energy and Water Resources.
- Federation of French Water Holding Companies.
- Organization for Information and Economic Forecasts of France.
- National Statistics Center Spain.

One of the optimal strategies to control and follow up on issues related to water resources management is paying attention to the frequency and level of access. The level of enjoyment of the countries of this group for sustainable water resources is not homogeneous in the different regions; the countries Romania, Germany, France, Norway, Denmark, Ireland, Luxembourg, Finland, England, Wales, Northern Ireland, and Scotland have abundant water resources. The countries of Belgium, Greece, Spain, Portugal, and Italy, especially in the regions in the South, have fewer water resources. England, USA and France have record levels of underground water resources; in 2011, the equivalent to 1444 million cubic meters were harvested. A decrease in the level of rainfall according to the geographical extent of continental European countries has been shown, and the volume of water evaporation is very high.

This value in European countries is between 14 and 94 cubic kilometers per year, which is different from other countries in the world. Considering the basic living conditions of the countries, India, China, USA, Pakistan, Japan, Thailand, Indonesia, Bangladesh, Mexico, and the Russian Federation are the biggest water users in the world. The volume of harvested water in most European countries is between 944 and 10,444 cubic meters per year. It was reported that between 294 and 944 cubic meters and between 144 and 294 cubic meters was harvested per year in England and Finland, respectively. The countries of Finland and Sweden have a record level of water resources per family due to low population density and growth than the countries of Germany, USA, and Poland. In the Czech Republic a lower level of more than 29% of the volume of stable water is available than most of the European countries. The water is consumed in agriculture. To reduce the volume of water loss in the agricultural sector, extensive activities are carried out. According to a report of recent years, water loss in the agricultural sector has been reduced by 34%. The consumption of consumers in these countries is at a favorable level and ranges from 92 to 224 liters per person. With the establishment of the European Union in late 1994, the Water Framework Directive was drafted. The strategies of the European Commission and the Economy Directive are these two comprehensive regional documents, including management.

In the integration of water resources, the emphasis is on the comprehensive management of wastewater systems from all aspects. The management of consumption and sustainable living and methods of waste management, production, and consumption have been the most important strategies for water and wastewater treatment in European countries. The basis of this directive is the programs and measures for the protection of water resources of the European Commission and the European Environmental Protection Agency.

The following items are related to the protection of water resources:

- Reduction of 11% of water consumption in the agricultural sector through productivity management and the use of new technologies.
- Increasing the quality level of water and reducing the amount of nitrates and other heavy metals.
- Improving the pattern of consumption, demand management, and optimal search for water consumption.
- Water-saving attitude and the establishment of water-saving efficiency systems.
- Hydrology management and protection of the river basin.
- Protection and maintenance of surface and underground water sources and health security of water sources.
- Hygienic disposal of waste from sewage treatment plants and the use of new technologies in the sanitary treatment of sewage.
- Cooperation of countries in the management and control of the exploitation of the common rivers.
- Balancing the amount of water withdrawal from rivers and catchment basins.
- Identification of water supply sources for areas that have water taps.
- Public education and information, especially the importance and sensitivity of regular visits to water and sewage.
- Coping with floods and natural disasters to protect water resources.

For 20 years, Germany has not had many problems in the field of water resource provision. Its preservation and maintenance for the future is one of the duties of the federal government. Currently one of the government's programs is the management and efficient use of a water resources framework. There are domestic, regional and international standards in this country. There are 12,000 installations and equipment for water supply and distribution, and thousand kilometers of water distribution networks have been built. More than 94 sources of drinking water are provided with microbiological methods.

According to the German constitution, the institutional sovereignty of municipalities in water and sewage management include the provision of drinking water distribution and sanitary waste disposal. There are duties of municipalities in the collection of water resources. Germany's annual renewal of rainfall and incoming water from neighboring countries is approximately 122 billion cubic meters. Currently, about 19% of these resources are used by applicants for agriculture, agriculture, and the general public. It is possible to spend 2.9% in the public sector, 2.6% in mining and renaissance, and 13.5% in thermal power plants. Public consumption is about 5.1 billion cubic meters per year, increasing public water consumption from 2.9 billion cubic meters in 2014 to 2.7.

The volume of unexploited water has increased from 81.0 million to 82.4 million in 2019. Germany's water consumption is 124 liters per person per day. Consumption depends on the amount of abundant water resources. Germany, in its policy with other EU countries, has managed water consumption and since the end of the 1924, with the approval of environmental regulations and water resources management, activities have been at the top of the list. Water consumption has been reduced by 19% [16–25].

Consumption management can refer to the following:

- Advertising, informing, and increasing the awareness of the community members in regular contact with water and sewage, so that, according to polls, more than 91% of the population of the country are aware of water reports regularly.
- The participants of training are mostly in elementary schools, so students have a theoretical aptitude of water and sewerage due in the educational rounds, according to the field observations of all water supply and distribution centers in particular. Water treatment plants are equipped with educational classes with modern educational equipment. Subscribers' behavior has the greatest impact on consumption management, so this topic is important in the context of behavioral economics.
- Extensive use of equipment to reduce water consumption in Germany from the point of view of production and distribution of equipment.
- Germany is one of the leading countries at the international level, and most of the public-city service centers have modern equipment. They are equipped with water-saving equipment, which causes a decrease in consumption.

- Economic communication: Water consumption management is reviewed in the context of theories and theoretical foundations of sustainable development and economy.

After China, Germany is the second country in the world to use the components of the rotational economy. This subject as a culture in society and federal statesmen has found a special place in practical and operational ways. The stability of this country is due to the management of water consumption, control of non-revenue water, the use of wastewater, and extensive experiments.

France, based on its area of 909,000 square kilometers, is the largest European country, and has a population of 19 million.

A report in 2015 showed that the annual rainfall of this country is 943 billion cubic meters, and it provides most of the 20,444 billion cubic meters of renewable water resources in France. The total volume of water withdrawn for urban and general consumption in France is 5.4 billion cubic meters. This is equivalent to 29 cubic meters per capita, and is equivalent to 230 liters per person per day. This volume of public water use in France is far less than the average of the European Union, and is much less than the average of other developed countries such as USA, Canada, China, Australia, and Japan. In the year 2011, the volume of harvested water for public use was 5.4 billion cubic meters. The urban use was around 5.5 billion cubic meters.

This estimate is based on two scenarios without considering climate change and global warming:

- Reducing the volume of water consumption for public-household purposes by applying operational strategies for water demand management.
- Reduction of water without income against water loss and costs, considering the valuable activities that have been carried out in this field.

One of the long-term strategies of the French Water Corporation is to reduce non-revenue water as much as possible according to the European Union directives.

The management of household consumption caused a decrease in the volume of water harvesting by 33% compared. It was according to the monitoring and statistics directorate. In 2017 behavioral changes in consumers were due to more knowledge of environmental issues, applying sustainable practices of water demand management of water companies. The improvement in the management of customers' consumption led to reduced water consumption. This was due to following factors:

- Advancement of research technologies in water consumption management and creating a culture of installing water consumption reduction devices.
- Government centers, schools, hospitals, etc.
- Changes in the structure of production and manufacturing of goods and providing services by reducing the volume of water consumption in the name of unification of production.

Increasing the efficiency of urban water distribution management systems occurs by the reduction and control of water management.

The volume of water consumption in France in 2012, based on the reports of the French National Institute of Statistics, was 102 liters.

In the management of water resources, meeting water demand depends on the optimal management of watersheds. There is an action plan for water management in USA as one of the leading countries in comparison with other European Union countries. The USA has always been fighting against water shortages due to regional geography and topography, and extensive strategies. In this context, the USA has also carried out water irrigation, since the beginning of 1914, especially in the southern areas.

In Europe, Spain has formulated and carried out special programs and strategies in terms of water supply, which have been very effective and efficient.

Successful programs of the government of Spain focused on the direction of water management. It includes the hydrological management of catchment basins, and non-transfer and displacement of water between watersheds. Demand management tools, and cultural programs to reduce consumption and increase consumption efficiency have been used. Agricultural water is provided by the use of desalination systems in the context of seawater desalination in this country [43–69].

This country is the most successful in the world in seawater treatment. It has carried out operational activities and expanded investments in this field. Based on a report in 2010, the amount of monsoon rainfall in Spain is equal to 3,040,444 square meters. It has been reported that this volume of precipitation is equivalent to 1,110,349 cubic meters of the circulation of annual flows. The water resources have been stable. In this regard, the storage volume of water reservoirs is 910,413 cubic meters. This is equivalent to 290,492 cubic meters of penetrated aquifers and 90,932 cubic meters were harvested for various purposes.

According to weather and crop harvest expected in 2021 to 2029, water production should increase. Also, climate change causes a decrease in the priority of water resources.

An increasing volume of consumer demand up to 14 times is one of the models of water management in many regions of Spain. The government is continuously following up and controlling, in addition to the above, social consequences, and life.

Over the last two decades, drought has become a frequent occurrence. Water is observed in almost all the catchment areas of this country and causes an increase. This phenomenon happens in the costs of harvesting special crops. The costs of the loss of farming and mechanization are based on the results of simulation model studies.

In Spain, with an increase of one degree in temperature affected by the main changes in the world and a decrease of 9% in annual precipitation, caused a decrease of 9%

in the amount of runoff water in the catchment basins, but the annual drought programs are very efficient.

It seems that the targeted policies of the water sector, issued to reduce the destructive effects of water loss, are efficient. The catchment basins are examples of the artificial provisions of the Spain government to fight against the main changes in water supply as described below:

- Comprehensive national programs to adapt to Islamic changes.
- The planning strategies of social and economic sectors have been targeted.
- Management regulations of watersheds and rivers have been introduced.

Climate change affects the management of water resources. Sustainable new programs of the government of Spain have been developed in the fight against water loss in the use of freshwater systems. These programs include wastewater reuse in the southern areas of the marginal cities of the Mediterranean Sea instead of transferring water between watersheds.

In this country, the water consumed in the public and domestic sectors is obtained from freshwater systems. It can be said that Spain has the first place in the European Union, based on the comprehensive report of the National

Institute of America. Utilization of surface water resources is 13% per year. The amount of withdrawal from underground water resources is 33% per year. Reverse osmosis water systems are mostly in the Mediterranean Sea and the southern part of the country. Water consumption in the country of Spain is comparable to some European and even international countries, at a relatively high level. The total of 129 liters per day per person in the domestic sector is low, from the point of view of this effective index. The consumption of each citizen to maintain health is 144 liters per day, according to the World Health Organization and Development Organization. Saving cooperation of 111 liters per day has been determined, and in this context, the situation of water consumption in Spain has been optimized. The volume of water supply distributed for domestic consumption in 2013 was equal to 19 million cubic meters of total water supply, and in the economic and public sectors, 199 million cubic meters, 21 million cubic meters, respectively. The municipality spends 292 million on irrigation of public parks, and washing of public roads and highways. The volume of water consumption in the year 2013 was 14 million per square meter. In the domestic consumption sector it was 3.9. The volume of consumption of each citizen of Spain in the year 2014 was reduced and was equivalent to 194 liters per days, which has been achieved with many activities; this pattern was reduced to 129 liters per day in 2019 [70–119].

In Spain, cultural and social targeted policies reduced the volume of water consumption at the level of European Union standards. Reducing water consumption is one of the strategic activities of the government of Spain and its units. The operation in the water and sewage system of Spain is programmed to reduce the per capita water consumption of consumers. It is benefit-taking the components of water demand management, especially on the cultural level. In this regard, according to the survey which was done by the citizens of Spain, most of the consumers are more than 92% of the consumption of European Union standards. In some areas of the south-eastern front and the coast of the Mediterranean Sea, they have continuous cooperation with the government. They have water and wastewater companies' activation in wastewater reuse.

In this work, the aim was to show that polyhydroxyalkanoates (PHAs) are a threat to wastewater transmission lines, and at the same time, due to the PHA production by wastewater reuse, this phenomenon can be raised as an opportunity and an important topic of sustainability and water-saving for design engineers [120–115].

CHAPTER

2

Methods

The smart control of wastewater supply based on remote sensing (RS) and the Internet of Things (IoT) can change the polyhydroxyalkanoates (PHAs) into an opportunity as an environmental resource. Wastewater reuse can be assumed to be a benefit for social security and access to economic considerations (Figures 2.1–2.3).

Figure 2.1: Wastewater reuse: aerobic, anaerobic tank, and aerobic reactor and a sedimentation tank.

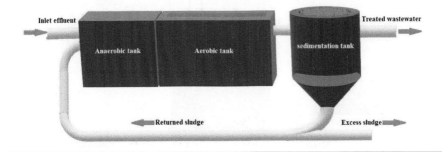

Figure 2.2: Wastewater reuse: anaerobic tank.

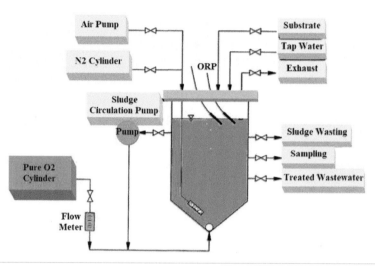

Figure 2.3: Wastewater reuse: anaerobic tank control system.

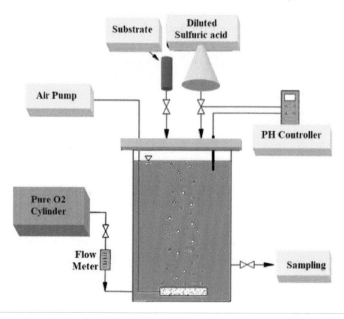

2.1 Fomulation

The fomulation of processes (Equations (2.1)–(2.12)) in anaerobic-aerobic activated sludge system and aeration tank the following:

$$V = LO/BV, \qquad (2.1)$$

$$BV = MLSS \times MLVSS, \qquad (2.2)$$

$$R_t = H/VS, \qquad (2.3)$$

$$R_t = V/Q. \qquad (2.4)$$

In the case of low *F/M*, micro-organisms feed on organic material in wastewater or feed on other micro-organisms.

In order to save the balanced condition, we need a high return sludge and high *MLSS*.

On the base of European standard, this happened for days without rain [28–31]. The ratio of *RS* per *max* influent become 100% and can be achieved from the following relations:

$$RS/Q_{\max} = [MLSS/(SMLSS - MLSS)] \times 100, \qquad (2.5)$$

There was a problem in the treatment process when the sludge volume index *S.V.I* became more than 200. In this condition, sedimentation failed. By decreasing *S.V.I*, the aeration time was decreased.

The amount of surplus sludge that must be removed from the settling tank was related to biochemical oxygen demand ≪ *BOD5* ≫ of influent entrance to the aeration tank and the settling tank output.

$$\text{Sludge age} = (MLSS) \times (V)/(SS_e \times Q + SLS), \qquad (2.6)$$

A portion of the absorbed pure oxygen spent gives the energy consumption and multiplying of bacteria. The other part of oxygen is spent on oxidation of organic carbonate and organic nitrogenous material.

$$Q_V = A_Y B_V + B(MLSS) + 3 - 4(ON), \qquad (2.7)$$

Process design:

$$O_C = [C_S/A(F)(C_S - C_X)] \times [.5YB_V + .1MLSS + 3.4ON], \qquad (2.8)$$

Design criteria:

Influent production per person = 200 (liters/day)

Max influent factor = 1.71 (per 14 hours)

$$BV = .5(KG.BOD5/M3DAY),$$
$$BV/MLSS = F/M = .15 \text{ and}$$
$$BOD5 = 60gr/p - d = .60kg/p - d,$$

Wastewater temperature = 20 °C.

In this work, pure oxygen was introduced into the wastewater by a 150 bar pressure and a 40 liter volume pure O_2 cylinder. Using submerged porous diffusers and air nozzles pure O_2 was introduced to the aeration tank. Then micro-organisms using the received pure O_2 grew quickly [116–143].

2.2 Research Approach

Pollution decreasing to 20 (mg/liter) for *BOD5*.

Influent pollution calculation:

- Fixed population = 700
- Office workers = 500
- Total population = 1200

$$Q_m^d = (700)(.2) + 500(.05) = 165 \, (m^3/day),$$

Max influent per hour:

$$Q^{dm} = 165/14 = 12 \, (m^3/h) = 3.3(\,L/s).$$

Total amount of influent pollution:

0.06(1200) = 72 (kg/day).

Average amount of influent pollution:

72/165 = 0.45 (kg/m^3) = 450 (mg/lit).

Due to pollution-detention curves, the decrease of influent pollution relation with detention time is:

$R_t = V/Q =$ tank volume/influent flow rate

$R_t + 200/12 = 17$ (h)

Pollution decreasing percent $= 0.35$

$450(1 - 0.35) = 293$ (mg/L).

Average amount of influent pollution:

(Per 24 hours) $= 0.293(165) = 48.5$ (kg BOD/day).

Oxygen required calculation (Figure 2.3):

$C_S = 9.17$ and $A = 0.9$

$Y = 0.925$ and $F = 0.85$

$MLSS = 3.3$ (kg/m^3)

$3.4(ON) = 0.23$

$$O_C = [C_S/A(F)(C_S - C_X)] \times [.5(Y)B_V + .1(MLSS) + 3.4ON], \quad (2.9)$$
$$O_C = [9.17/.9(.85)(9.17 - 1.5)] \times [.5(.925).5 + .1(3.3) + 3.4(ON)], \quad (2.10)$$
$$O_C = 1.2366 \ (\text{kgO}_2/\text{m}^3 \text{ day}), \quad (2.11)$$
$$\sum O_C = [200(1.2366)]/24 = 10.3 \ (\text{kgO}_2/\text{h}) \quad (2.12)$$

CHAPTER 3

Results and Discussion

The results of this work through pure O_2 injection in a conventional activated sludge proccess for PHA production in wastewater are shown in Figures 3.1–3.10.

Aeration tank volume $= 200 (M3)$

Detention time: $R_t = V/Q = 200/12 = 17$ (h)

Theoretical required volume: $V = Lo/BV = 72/.5 = 144 \text{ m}^3$

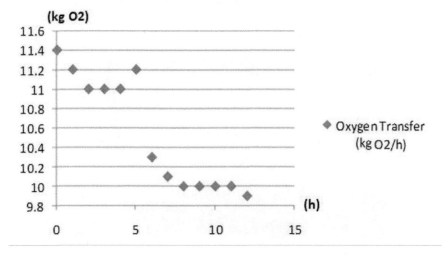

Figure 3.1: Wastewater reuse: pure oxygen required for production of PHA.

Figure 3.2: Wastewater reuse: pure oxygen introduced in wastewater.

Figure 3.3: Wastewater reuse: pure oxygen gas station.

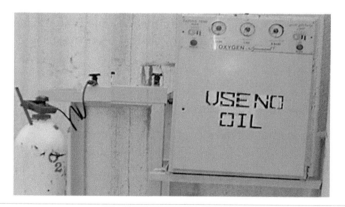

Figure 3.4: Wastewater reuse: pure oxygen injection to wastewater.

Figure 3.5: Wastewater reuse: aeration basin.

Figure 3.6: Wastewater reuse: sedimentation tank.

Figure 3.7: Wastewater reuse: aeration tank.

Figure 3.8: Wastewater reuse: deep aeration.

Figure 3.9: Wastewater reuse: conventional activated sludge lagoon.

Results and Discussion

Figure 3.10: Wastewater reuse: aerobic and anaerobic tank.

Theoretical detention time:

$R_t = V/Q = 144/12 = 12$ (h).

3.1 Regression and Wastewater Reuse

In this work, there was a linear relation between the variables of research. The scatter diagram gave the basic idea. The regression equations were used to draw the regression line.

3.2 Regression Variables; PHA and Reference Values

Based on the main results of this work, for production of PHA there was a significant relation through the power function between the PHA and the reference value in the production process of PHA (Equations (3.1)–(3.2)).

$$\text{PHA} = f(\text{Reference}), \tag{3.1}$$
$$\text{PHA} = \text{dependent variable} : \text{PHA}(\%).$$
$$\text{H} = \text{independent variable} : \text{reference}.$$
$$f(x) = y = cx^P \tag{3.2}$$
$$y = 21768.626 x^{-1.216}$$

where:

y = PHA
x = references value.

The results of this work showed that there was a significant relation between PHA and the reference value. By regression analysis and through investigation of the scatter diagram and the curve fit due to regression analysis, the PHA against the reference value Equation (3.1) showed a good correlation between the dependent variable (PHA) and the independent variable (reference value) data.

The data collection system includes an anaerobic tank, aerobic reactor, sedimentation tank and monitoring equipment. These data were used for regression analysis. The data were detected by networked sensors which were installed for measuring the PHA and reference parameters in the wastewater reuse system. The PHA and reference values also were intercommunicated by remote sensing (RS) and Internet of Things (IoT). The data of PHA and reference values were transmitted to the control unit of the SBR system (Table 3.1).

Table 3.1: Wastewater reuse: data collection results.

No.	PHA (%)	Reference
1	88.2	98
2	97.1	97
3	87.3	98
4	77.2	98.2
5	65.2	99
6	87.5	97
7	88.4	89

3.3 Regression Analysis

In regression analysis the regression lines reflected the overall movement path of scattered points in the nominal coordinate system (Figures 3.11–3.17). The regression lines showed the intensity and weakness and the type of correlation

Figure 3.11: Regression analysis: scatter diagram.

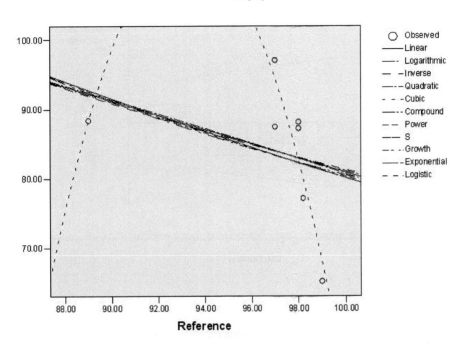

Figure 3.12: Regression analysis: scatter diagram.

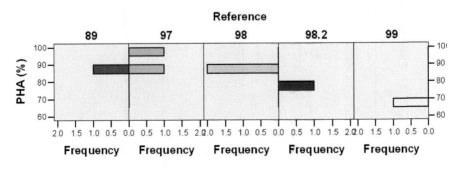

Figure 3.13: Regression analysis: scatter diagram.

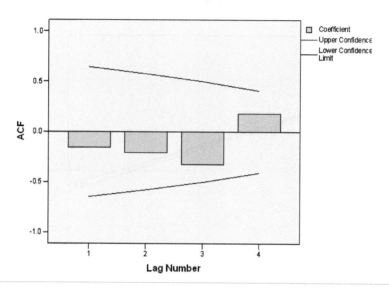

Figure 3.14: Regression analysis: scatter diagram.

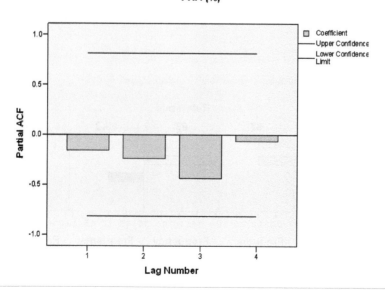

Figure 3.15: Regression analysis: scatter diagram.

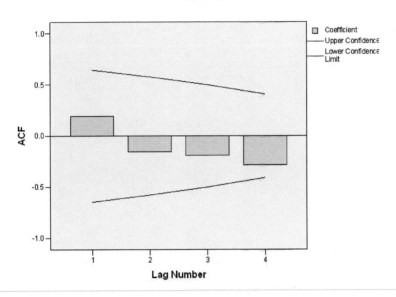

Figure 3.16: Regression analysis: scatter diagram.

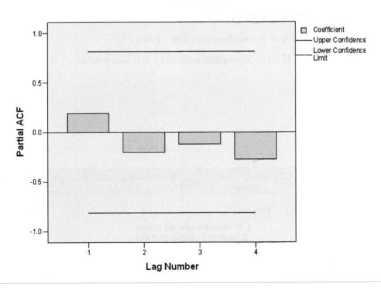

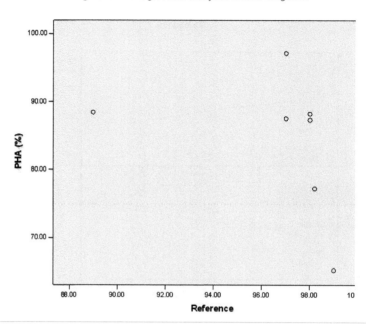

Figure 3.17: Regression analysis: scatter diagram.

between the variables including the data of PHA (Equation (3.1)) and reference values (Tables 3.2–3.10).

$$\text{PHA} = \text{f(Reference)}, \tag{3.3}$$
$$\text{PHA} = \text{dependent variable}: \text{PHA}(\%)$$
$$\text{H} = \text{independent variable}: \text{reference value}.$$

Table 3.2: Wastewater reuse: data collection results; variables entered/removed.

Model	Variables Entered	Variables Removed	Method
1	Reference(a)	.	Enter

a All requested variables entered.
b Dependent Variable: PHA (%)

Table 3.3: Wastewater reuse: data collection results; model summary.

Model	R	R Square	Adjusted R Square	Std. Error of the Estimate
1	.343(a)	.118	-.058	10.54412

a Predictors: (Constant), Reference

Table 3.4: Wastewater reuse: data collection results; model description.

Model Name		MOD_1
Dependent Variable	1	PHA (%)
Equation	1	Linear
	2	Logarithmic
	3	Inverse
	4	Quadratic
	5	Cubic
	6	Compound(a)
	7	Power(a)
	8	S(a)
	9	Growth(a)
	10	Exponential(a)
	11	Logistic(a)
Independent Variable		Reference
Constant		Included
Variable Whose Values Label Observations in Plots		Unspecified
Tolerance for Entering Terms in Equations		.0001

a The model requires all non-missing values to be positive.

Results and Discussion

Table 3.5: Wastewater reuse: data collection results; model summary and parameter estimates.

Dependent Variable: PHA (%)

Equation	Model Summary					Parameter Estimates			
	R Square	F	df1	df2	Sig.	Constant	b1	b2	b3
Linear	.118	.669	1	5	.451	183.734	-1.028		
Logarithmic	.113	.635	1	5	.462	514.231	-94.051		
Inverse	.108	.604	1	5	.472	-4.677	8596.308		
Quadratic	.123	.704	1	5	.440	136.867	.000	-.006	
Cubic	.805	8.261	2	4	.038	-7835.762	127.785	.000	-.005
Compound	.124	.706	1	5	.439	302.698	.987		
Power	.118	.671	1	5	.450	21768.626	-1.216		
S	.113	.639	1	5	.460	3.276	111.254		
Growth	.124	.706	1	5	.439	5.713	-.013		
Exponential	.124	.706	1	5	.439	302.698	-.013		
Logistic	.124	.706	1	5	.439	.003	1.013		

The independent variable is Reference.

Table 3.6: Wastewater reuse: data collection results.

Likelihood Ratio Tests

Effect	Model Fitting Criteria	Likelihood Ratio Tests		
	-2 Log Likelihood of Reduced Model	Chi-Square	df	Sig.
Intercept	2.773(a)	.000	0	.
VAR00002	24.470	21.698	24	.597

The chi-square statistic is the difference in -2 log-likelihoods between the final model and a reduced model. The reduced model is formed by omitting an effect from the final model. The null hypothesis is that all parameters of that effect are 0.
a This reduced model is equivalent to the final model because omitting the effect does not increase the degrees of freedom.

Table 3.7: Wastewater reuse: data collection results; autocorrelations.

Series: PHA (%)

Lag	Autocorrelation	Std. Error (a)	Box-Ljung Statistic		
			Value	df	Sig. (b)
1	-.153	.323	.224	1	.636
2	-.210	.289	.751	2	.687
3	-.323	.250	2.425	3	.489
4	.180	.204	3.199	4	.525

a The underlying process assumed is independence (white noise).
b Based on the asymptotic chi-square approximation.

Table 3.8: Wastewater reuse: data collection results; partial autocorrelations.

Series: PHA (%)

Lag	Partial Autocorrelation	Std. Error
1	-.153	.408
2	-.239	.408
3	-.434	.408
4	-.065	.408

Table 3.9: Wastewater reuse: data collection results; autocorrelations.

Series: Reference

Lag	Autocorrelation	Std. Error (a)	Box-Ljung Statistic		
			Value	df	Sig.(b)
1	.193	.323	.357	1	.550
2	-.157	.289	.653	2	.721
3	-.189	.250	1.227	3	.747
4	-.285	.204	3.172	4	.529

a The underlying process assumed is independence (white noise).
b Based on the asymptotic chi-square approximation.

Table 3.10: Wastewater reuse: data collection results; autocorrelations.

Series: Reference

Lag	Partial Autocorrelation	Std. Error
1	.193	.408
2	-.202	.408
3	-.123	.408
4	-.274	.408

3.4 Wastewater Reuse System Control

The wastewater reuse system control in the present work is shown in the schematic of the polymer production system (Figure 3.18). All the steps are precisely controlled by a closed loop control system (CLCS). In this system, the intensity of oxidation and reduction reactions in the aerobic and anaerobic stages will be measured and controlled by an oxidation reduction potential (ORP) controller. In the field method, the data is received from pressure gauges and flowmeters with a remote reading. The SBR system used in this work is shown in Figure 2.3. The SBR system was equipped with remote sensing (RS) and Internet of Things (IoT).

Figure 3.18: Wastewater reuse: closed loop control system (CLCS).

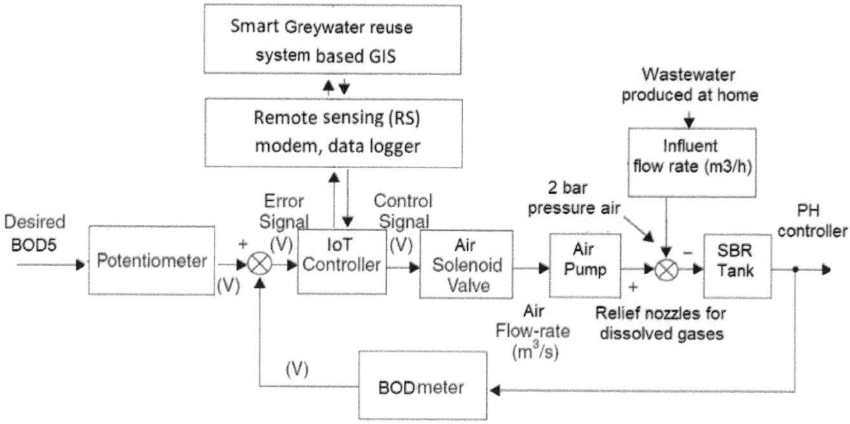

The data intercommunication facilities were installed in a polymer production system. These facilities included an anaerobic tank, aerobic reactor, sedimentation tank, and parameter monitoring equipment. They were installed for measuring the PHA and reference parameters in the polymer production system. These facilities detected the PHA and reference values by remote sensing (RS) and Internet of Things (IoT).

CHAPTER

4

Conclusion

Wastewater reuse can lead to the degradable plastics that have been prepared so far, a group called polyhydroxyalkanoates (PHAs) and its copolymers as biodegradable plastic. In wastewater reuse, biodegradable plastics have been given more attention. The PHAs are polyester hydroxyalkanoates and their molecular weight depends on the type of micro-organism and growth conditions. These types of compounds are used even in surgical operations due to their physical and mechanical properties and also due to their compatibility with the body. The results of this work show that urban wastewater can be used as a cheap source in the production of greywater.

This work showed that in the wastewater reuse process can lead to degradable plastics. This plastic also is called bio-plastic. By regression analysis and through investigation of the scatter diagram and the curve fit due to regression analysis, PHA against the reference value equation showed a good correlation between the dependent variable (PHA) and the independent variable (reference value) data.

References

[1] Adline S.M. Chua, Hiroo Takabatake Hiroyasu Satoh and Takashi Mino. 2003 Production of polyhydroxyalkanoayes (PHA) by activated sludge treating municipal wastewater: effect of pH, sludge retention time (SRT), and acetate concentration in influent. Wat. Res. 37: 3602-3611

[2] Asli H. H., Arabani M. and Golpour Y. (2020) Reclaimed asphalt pavement (RAP) based on a geospatial information system (GIS). *Slovak Journal of Civil Engineering*, Slovak University of Technology in Bratislava, Slovak, Vol. 28, Issue 2, 36-42, https://doi.org/10.2478/sjce-2020-0013.

[3] Asli H. K., Khodaparast Haghi A., and Asli H. H., Sabermaash Eshghi E. (2012). Water Hammer Modelling and Simulation by GIS, *Modelling and Simulation in Engineering*, Hindawi Publishing Corporation, Volume 2012, Article ID 704163, 4 pages, doi:10.1155/2012/704163, ISSN: 16875605, 16875591, Pub., Vol.1, No. 3, USA, http://www.hindawi.com/journals/mse/aip/704163/.

[4] Asli H. H. and Arabani M. (2022). Analysis of Strain and Failure of Asphalt Pavement, Computational Research Progress in Applied Science & Engineering, *Transactions of Civil and Environmental Engineering* 8, 1–11, Article ID: 2250. https://doi.org/10.52547/crpase.8.1.2250.

[5] Akiyama M, Taima YM Doi Y. 1992. Production of PHB by a bacterium of the Genus alcaligesesis utilizing long chain fatty acids. Appl. Microbiol. Biotechnol, 37:689-701.

[6] Akiyama M, Doi Y. 1993. Production of PHB from alkanedioic Acids and hydroxylated fatty acids by Alcaligesesis sp. Biotechnol. Letter. 15(2)163-168.

[7] Anderson, A.J. and Dawes, E.A. (1990). Occurrence, metabolism, metabolic role, and industrial uses of bacterial polyhydroxyalkanoates. Microbiological Reviews.54 (4),450-472.

References

[8] Anderson, A.J., Haywood, G.W., Williams, D.R., and Dawes, E.A. (1990). The production of polyhydroxyalkanoates from unrelated carbon sources. In: Novel biodegradable microbial polymers, E. A. Dawes (ed.), Kluwer Academic Publishers, The Netherlands, 119-129.

[9] Hariri Asli H. (2022) Investigation of the Factors Affecting Pedestrian Accidents in Urban Roundabouts, Computational Research Progress in Applied Science & Engineering (CRPASE), Article ID: 2255,8(1), 1–4, https://crpase.com/archive/CRPASE-Vol-08-issue-01-15172135.pdf, https://doi.org/10.52547/crpase.8.1.2255.

[10] Hossein Hariri Asli Ali Pourhashemi, Ann Rose Abraham, A. K. Haghi (2023) New Advances in Materials Technologies; Experimental Characterizations, Theoretical Modeling, and Field Practices , Hard ISBN: 9781774914847, https://www.appleacademicpress.com/new-advances-in-material-technologies-experimental-characterizations-theoretical-modeling-and-field-practices/9781774914847

[11] Hossein Hariri Asli, Tamara Tatrishvili, Ann Rose Abraham, A. K. Haghi, Sustainable Water Treatment and Ecosystem Protection Strategies, Hard ISBN: 9781774915189, https://www.appleacademicpress.com/sustainable-water-treatment-and-ecosystem-protection-strategies-/9781774915189\protect \protect\relax$\@@underline{\hbox{\protect\setbox\tw@\hbox{\begingroup\pdfcolorstack\main@pdfcolorstackpush{0gOG}\aftergroup\pdfcolorstack\main@pdfcolorstackpop\relax \endgraf\endgroup}\dp\tw@\z@\box\tw@}}\mathsurround\z@$\relax

[12] Anderson, A.J., Williams, D.R., Taidi, B., Dawes, E.A., and Ewing, D.F. (1992). Studies in copolyester synthesis by Rhodococcus rubber and factors influencing the molecular mass of polyhydroxybutyrate accumulated by Methylobacterium extorquens and Alcaligenes eutrophus. FEMS Microbiology reviews. 130,93-102.

[13] Ayorinde FO, Saeed KA, Eribo E, Morrow A, Collis WE, McInnis F, Polack SK, Eribo Be. 1998. Production of PHB from saponified veronica galamesis oil by Alcaligenesis eutrophus. J. Indus. Microbiol & Biotechnol, 21:46-50.

[14] Berger, E.; Ramsay, B.A.; Ramsay, J.A. and Chavarie, C., "PHB recovery by hypochlorite digestion of non-PHB biomass", Biotech. Techniq., 1989, 3(4), 227-232.

[15] Booma, M., Selke, S.E., and Giacin, J.R. (1994). Degradable plastics. Journal of Elastomers and Plastics. 26, 104-142.

[16] Bourque, Ouellette, B., Andre, G., and Groleau, D. (1992). Production of poly-hydroxybutyrate from methanol: Characterization of a new isolate of Methylobacteruim extorquens. Appl. Microbiol. Biotechnol. 37, 7-12.

[17] Bourque, D., Pomerleau, Y., and Groleau, D. (1995). High cell density production of poly- hydroxdybutyrate (PHB) form methanol

by Methylobacteruim extorquens: Production of high-molecular-mass PHB. Appl. Microbiol. Biotechonol. 44,367-376.

[18] Brandl, H.; Gross, R.A.; Lenz, R.W. and Fuller, R.C., "Plastics form bacteria and for bacteria: poly (hydroxybutyrate) as natural, biodegradable polyesters", Advances in Biochemical Engineering, 1990,41,77-93.

[19] Brandl, H., Bachofen, B., Mayer, J., and Wintermantel, E. (1995). Degradation and applications of polyhydroxyalkanoates. Can. J. Microbiol. 41(suppl.1), 143-153.

[20] Braunegg, G.; Sonnleitner, B and Lafferty, R.M., "A rapid gas chromatographic method for the determination of PHB in microbial biomass", Eur. J. Appl. Microb. Biotech.,1978, 6: 905-910.

[21] Braunegg, G., Lefebvere, G., and Genser, K.F. (1998). Polyalkanoates, biopolyester from renewable resources: Physiological and engineering aspects. Journal of Biotechnology. 65,127-161.

[22] Byrom, D. (1987). Polymer synthesis by microorganisms: Technology and economics. TIBTECH. 5,246-250.

[23] Byrom, D. (1990). Industrial production of Copolymer from Alcaligenes eutrophus. In: Novel biodegradable microbial polymers, Dawes. E.A. (ed.), Kluwer Academic Pubishers, the Netherlands, 113-117.

[24] Byrom, D. (1992). Production of poly-hydroxybutyrate: poly-hydroxyvalerate copolymers. FEMS Microbiology Reviews. 103,247-250.

[25] Castella JM, Urmenta J, Lafuente R, Navarrte A, Guerrero R. 1995. Biodegradation of PHAs in aerobic sediments. Inter. Biodeter. Biodegrad., 155-174.

[26] Doi, Y. (1990). Microbial Polyesters. VCH Publishers, Inc. New york, USA.

[27] Doi Y., Segawa, A., Nakamura, S., and kunioka, M. (1990). Production of biodegradable copolymers by Alcaligenes eutrophus. In: Novel biodegradable microbial polymers, Dawes, E.A. (ed.), kluwer Academic Pubishers, The Netherlands, 37-48.

[28] Fidler, S. and dennis, D. (1992). Polyhydroxyalkanoate production in recombinant Escherichia coli. FEMS Microbiology Reviews. 103,231-236.

[29] Fuchtenbusch B, Steinbuchel A. 1999. Biosynthesis of PHAs from low rank coal liquefaction products by Pseudomonas oleovorance and Rhodococcus rubber, Appl. Microbiol. Biotechnol, 52:91-95.

[30] Fuchtenbusch B, Wullbrandt D, Steinbuchel A. 2000. Production of PHAs by Ralstonia eutroph and Pseudomonas oleovorance from oil remaining from biotechnological Rhamnose production. Appl. Microbiol. Biotechnol, 53:167-172.

References

[31] Gorenflo, V.; Steinbuchel, A.; Marose, S.; Rieseberg, M. and Scheper, T., "quantification of bacterial polyhydroxyalkanoic acids by Nile red staining", Appl. Microbiol. Biotechnol., 1999, 51,765-772.

[32] Griffin, G.J.L., "Chemistry and Technology of Biodegradable Polymers", 1994, Chapman & Hall, 1st Ed.

[33] Gross RA, DeMello C, Lenz RW, Brandl H, Fuller C. 1989. Biosynthesis and characterization of PHAs produced by Pseudomnas oleovorance, Macromolecules: 22:1106-1115.

[34] Grothe E, Chisti Y. 2000. PHB thermoplastic production by Alcaligenesis latus behavior of fed-batch Culture, 22:441-449.

[35] Hahn, S.K., Chang, Y.L., Kim, B.S., and Chang, H.N. (1994). Optimization of microbial poly (3-hydroxybutyrate) recovery using dispersions of sodium hypochlorite solution and chloroform. Biotechnology and Bioengineering. 44,250-262.

[36] Harrison, S.; Dennis, J.S. and Chase, S.A., "The effect of culture history on the distruption of Alcaligenes eutrophus by high pressure homogenization". In: Separation for Biotechnology II, Edited by D.L. Pyle, London, Elsvier, 1990, 1st Ed.

[37] Hassan, Mohd.Ali and et.al. (1997), The production of polyhydroxyalkanoate form anaerobically treated palm oil mill effluent by Rhodobacter sphaeroides, J. of. Rermentatioon and Bioengineering, 83(5), 485-488.

[38] Haywood, G. W., Anderson, A.J., and Dawes, E.A. (1989). A survey of the accumulation of novel polyhydroxyalkanoates by bacteria. Biotechnology Letters. 11(7), 471-476.

[39] Haywood, G.W., Anderson, A.J., Williams, D.R., and Dawes, E.A. (1991). Accumulation of a poly (hydroxyalkanoate) copolymer containing primarily 3-Hydroxyvarate from simple carbohydrate substrates by Rhodococcus sp. NCIMB 40126. Internatioanl Journal of Biological Macromolecules. 13,83-88.

[40] Hejazi, P.; Vasheghani-Farahani, E.; Yamini, Y. Supercritical Fluid Disruption of Ralstonia eutropha for Poly (hydroxybutirate) Recovery. Biotechnol. Prog. 2003, 19, 1519-1523.

[41] Hezayen FF, Rehm BHA, Eberhardt R. 2000. Polymer production by 2 newly isolation extremely halophilic archaea: Application of a Novel Corrosion Resistance Bioreactor.

[42] Holmes, P.A. (1985). Application of PHB-a microbially produced biodegradable thermoplastics. Phys. Technol. 16,32-36.

[43] Hood,C.R. and Randall, A.A. (in press). A biochemical hypothesis explaining the response of enhanced biological phosphorus removal biomss to organic substrates, Water Research.

[44] Horib K, Marsudi S, Unno H. 2002. Simulatanous production of PHAs and Rhamnolipids by Pseudomonas aeroginosa, Biotechnol. Bioeng., 78(6):699-707.

[45] Hrabak, O. (1992). Industrial production of poly-☐-hydroxybutyrate. FEMS Microbiology Reviews. 103,251-256.

[46] Huang, J., Shetty, A.S., and wang, M. (1990). Biodegradable plastics: A Review. Advances in Polymer Technology, 10(1), 23-30.

[47] Ishihara Y., Shimizu H., Shioya S. 1994. Mole fraction of PHB in fed-batch culture of A.eutrophus, Biotechnol. Bioeng., 81(5): 422-428.

[48] Jan, S.; Roblot, C.; Courtois, B.; Babotin, J.N. and Seguin, J.P., "H NMR spectroscopic determination of poly-hydroxybutyrate extracted from microbial biomass", Enz. Microb. Technol, 1996, 18,195-201.

[49] Asli, H. H., Arabani, M. and Golpour, Y. 2020 Reclaimed asphalt pavement (RAP) based on a geospatial information system (GIS). Slovak Journal of Civil Engineering. Slovak University of Technology in Bratislava. Slovak. 28(2). 36-42. https://doi.org/10.2478/sjce-2020-0013.

[50] Hariri Asli, K., Hariri Asli, H., Motlaghzadeh K., & Hariri Asli K. (2013). Numerical Techniques in Water Transmission, Frontiers of Engineering Mechanics Research(FEMR), Aug., Vol. 2 Iss. 3, PP. 56-62, ISSN: 2306-6016 (Online), ISSN: 2306-6024(Print), published by the world academic publishing co., limited, Hong Kong, Corpus ID: 108917427, http://www.acadÂemicpub.org/femr/, https://api.semanticscholar.org/CorpusID:108917427

[51] Hariri Asli, Kian, and Hariri Asli, Kaveh. 2022 Isolated pressure zones based on GIS as a solution for water network problems. Water Practice and Technology. wpt2022119. doi: https://doi.org/10.2166/wpt.2022.119.https://iwaponline.com/wpt/article/doi/

[52] Jian.yu, yingtao.Si, Wan.Keung R.Wong, (2002), Kinetics modeling of inhibition and utilization of mixed volatile fatty acids in the formation of polyhydroxyalkanoates by Ralstonia eutropha., Process Biochemistry, 37,731-738.

[53] Kato N, Konishi H, Shimo M, Sakazawa C. 1992. Production of PHB trimmer by Bacillus megaterium B124, J. Fermen. Bioeng., 73(3):246-247.

[54] Kellerhals MB, Hazenberg W, Witholt B. 1999. High cell density fermentation of Pseudomonas oleovorance for the production of mcl-PHA in two liquid phase media. Enz. Microbial Technol., 24:111-116.

[55] Khatipov E, Miyake M, Miyake J, Asads Y. 1998. Accumulation of PHB by Rhodobacter sphearoides on variouse carbon source and Nitrogen substrates, FEMS Microbiol letter, 1652:39-45.

[56] Khosravi-Darani, K.; Vasheghani-Farahani, E.; Shojaosadati, S.A.; Yamini, Y. Effect of Process Variables on Supercritical Fluid Disruption

References

[57] of Ralstonia eutropha Cells for poly (R- hydroxybutirate) Recovery. Biotechnol. Prog. 2004, 20, 1757-1765.

Kim BS, Lee Sc, Lee SY, Chang HN. 1994. Production of PHB by fed-batch culture of Ralstonia eutropha with glucose concentration control, Biotecnol. Bioeng, 43: 892-898.

[58] Kimura H, Yoshida Y, Doi Y. 1992. Production of PHB by Pseudomonas acidovorance, Biotech. Lett. 14(6):149-158.

[59] Kulaev, I.S. and Vagabov, V.M. (1983). Polyphosphate metabolism in microorganisms. Adv. Microbial physiol. 24,83-171.

[60] Hossein Hariri Asli, Tamara Tatrishvili, Ann Rose Abraham, A. K. Haghi, Sustainable Water Treatment and Ecosystem Protection Strategies, Hard ISBN: 9781774915189, https://www.appleacademicpress.com/sustainable-water-treatment-and-ecosystem-protection-strategies-/9781774915189

[61] Kian Hariri Asli (2024) Technological Advancement in HVAC&R System Energy Efficiency, Technological Advancement in Clean Energy Production, (Apple Academic Press, Hard ISBN: 9781774915585), https://www.appleacademicpress.com/title.php?id=1416

[62] Kian Hariri Asli (2024) IoT-Based Smart Water Leak Detection System for a Sustainable Future: A Case Study, Sustainable Water Engineering; Smart and Emerging Technologies, (Apple Academic Press, Hard ISBN: 9781774915714), https://appleacademicpress.com/sustainable-water-engineering-smart-and-emerging-technologies/9781774915714

[63] Kumagai, Y. (1992). Enzymatic degradation of binary blends of microbial poly (3-Hydroxy butyrate) with enzymatically active polymers, Polym. Degrad. Stab., 37,253-256.

[64] Lafferty, R.M., Korsatko, B., and korsatko, w. (1988). Microbial production of poly-hydroxybutyric acid. In: Biotechnology, Rehm, H.J. and Reed, G. (eds.). VCH publishers, New York, 135-176.

[65] Law, J. and Slepecky, R., "Assay of poly-hydroxybutyric acid", J. Bacteriol.,1961,82,33-36.

[66] Lee, S.Y. (1996). Review bacterial polyhydroxyalkanoates. Biotechnol. and Bioeng. 49,1-14.

[67] Lee S. and Yu, J. (1997). Production of biodegradable thermoplastics from municipal sludge by a two-stage bioprocess. Resources, Conservation and Recycling, 19,151-164.

[68] Lee, Sang.yup, Choi. Jong-il. (1999), Production and degradation of polyhydroxyalkanoates in waste environment, Waste Management, 19(2), 133-139.

[69] Lemos P.C., et.al. (1998), Effect of carbon source on the formation of polyhydroxyalkanoates (PHA) by a phosphate-accumulation mixed culture, Enzyme and Microbial Techonology, 22,662-671.

[70] Libergesel M, Husted E, Timm A, Steinbuchel A, Fuller R.C, Len ZRW and Schlegel HG. (1991), Formation of PHA by phototrophic and chemulithotrophic bacteria. Arch Microbial. 155: 415-421.

[71] Ling, Y; wong, H.H.; Thomas, C.J.; Williams, D.R.G. and Middelberg, A. P.J., "Pilot-scale extraction of PHB from recombinant E. coli by homogenization and centrifugation", Biosepatation, 1997,7,9-15.

[72] Mino, T., Kawakmi, T., and Matsuo, T. (1985a). Location of phosphorus in activated sludge and fraction of intracellular polyphosphate in biological phosphorus removal process. Wat. Sci. Technol. 17(2/3), 93-106.

[73] Pelissero, A. (1987). Update on Biodegradable Plastics Materials. Imballaggio, 38,54.

[74] Pfeffer, J.T. (1992). Recycling. Solide Waste Managt. Eng., 72-84.

[75] Preiss, J. (1984). Bacterial glycogen synthesis and its regulation. Ann. Rev. Microbiol. 38, 419-458.

[76] Preusting H, Kingama J, Witholt B. 1991. Physiology and polyester formation of pseudomonas oleovorance in continuous two liquid phase cultures. Enzy. Microbial Technol 13:770-780.

[77] Preusting H, Hazenberg W. 1993. Continuous production of PHB by pesudomonas oleovorance in a high cell; density, two liquid phases Chemostat, Enz. Microbial Technol., 15:311-316.

[78] Ramsay, J.A., Berger, E., Voyer, C., Chavarie, C., and Ramsay, B.A. (1990). Extraction of poly-3-hydroxybutyrate using chlorinated solvents. Biotechnol. Techninq. 8(8), 589-594.

[79] Ramsay, J.A., Berger, E., Ramsay, B.A., and Chavarie, C. (1994). Recovery of poly-3- hydroxyalkanoic acid granules by a surfactant-hypochlorite treatment. Biotechnol. Techniq.9(10), 709-712.

[80] Ramsay, J.A.; Berger, E.; Voyer, R.; Chavarie, C. and Ramsay, B.A., "Extraction of poly-3-hydroxybutyrate using chlorinated solvents", Biotech. Techniq., 1994, 8(8), 589-594.

[81] Ramsay, J.A.; Berger, E.; Ramsay, B.A. and Chavarie, C., "Recovery of poly-hydroxyalkanoic acid granules by a surfactant-hypochlorite treatment", Biotechnol. Techniq., 1990, 4(4), 221-226.

[82] Asli KiH, Asli KaH. "Smart Water System and Internet of Things." J Mod Ind Manuf, 2023; 2: 5. DOI: 10.53964/jmim.2023005, https://www.innovationforever.com/article.jmim20230111

[83] Asli, H. H. 2023 Modeling of Corrosion for Water System by Networked Sensors and the Internet of Things (IoT) in Compliance with Geography Information System (GIS). Sustainable Water Treatment and Ecosystem Protection Strategies. Hard ISBN: 9781774915189. https://www.appleacademicpress.com/sustainable-water-treatment-and-ecosystem-protection-strategies-/9781774915189.

References

[84] Renner, G., Haage, G., and Braunegg, G. (1996). Production of short-side-chain polyhydroxyalkanoates by various bacteria from the rRNA superfamily III. Appl. Microbiol. Biotechnol. 46,268-272.

[85] Resch, S., Gruber, K., Wanner, G., Slater, S., Dennnis, D., and Lubitz, W. (1998). Aqueous release and purification of poly (Hydroxybutyrate) from Escherichia coli. Journal of Biotechnology. 65, 173-182.

[86] Roh, K.S., Yeom, S.H., and Yoo, Y.J. (1995). The effects of sodium bisulfate in extraction of PHB by hypochlorite, Biotechnology Techniques. 4(4), 221-226.

[87] Samuel Lee, Jian yu. (1997), Production of biodegradable thermoplastics from municipal sludge by a two stage bioprocess, Resources, Conservation and Recycling, 19,151-164.

[88] Sasikala, CH. And Ramana, CH.V. (1996). Biodegradable polyesters, In: Advances in Applied Microbiology, Volume 42, Neidleman, S.L. and Laskin, A.I. (eds.), Academic press, California, 97-218.

[89] Satoh, H., Mino, T., and Matsuo, T. (1992). Uptake of organic substrates and accumulation of polyhydroxyalkanoates liked with glycolysis of intracellular carbohydrates under anaerobic conditions in the biological excess phosphorus removal processes, Wat. Sci. Tech. 26, 5-6, 933-942.

[90] Satoh, H., Iwamoto, Y., Mino, T., and Matsuo, T. (1998). Activated sludge as a possible source of biodegradable plastic. Proceedings, Water Quality International 1998, Book 3 Wastewater Treatments, pp. 304-311. IAWQ 19th Biennial International Conference, Vancouver, BC, Canada, 21-26 June 1998.

[91] Schwien U, Schmidt E. 1982. Improved degradation of monochlorophenol by constructed strain. Appl. Environ. Microbial., 44:33-39.

[92] Asli H. H. 2023 Applications of Networked Sensors and the Internet of Things (IoT) for Water Treatment. Sustainable Water Treatment. and Ecosystem Protection Strategies. Hard ISBN: 9781774915189.https://www.appleacademicpress.com/sustainable-water-treatment-and-ecosystem-protection-strategies-/9781774915189.

[93] Asli KiH, Asli KaH. "Smart Heating, Ventilating, Air-conditioning and Refrigeration by Web-based Geographic Information System". J Mod Ind Manuf, 2023; 2: 6. DOI: 10.53964/jmim.2023006, https://www.innovationforever.com/article.jmim20230139.

[94] Kian Hariri Asli, Kaveh Hariri Asli, Sajad Nazari; "Computational fluid dynamics analysis for smart control of water supply". Water Supply 2023; ws2023306. doi: https://doi.org/10.2166/ws.2023.306

[95] Senior, P.J., Beech, G.A., Ritchie, G.A.F., and Dawes, E.A. (1972). The role of oxygen limitation in the formation of poly-hydroxybutyrate

during batch and continuous culture of Azotobacter beijerinckii. The Biochemistry Journal. 128,1193-1201.

[96] Shimizu H, Tamira S, Shioya S, Suga K. 1993. Kinetics study of PHB productionjn and its molecular weight distribution control in fed-batch culture of A. eutrophus, J. Fermenta. Bioeng. 76(6)465-469.

[97] Shimizu H, Tamira S, Shioya S, Suga K. 1993. Kinetics study of PHB productionjn and its molecular weight distribution control in fed-batch culture of A. eutrophus, J. Fermenta. Bioeng. 76(6)465-469.

[98] Shimizu, H., Tamura, S., Ishihara, Y., Shioya, S., and Suga, k. (1994). Control of molecular weight distribution and mole fraction in Poly (D 3hydroxyalkanoates) (PHA) production by Alcaligenes eutrohpus. In: Biodegradable plastics and polymers, Doi, Y. and fukuda, K. (eds.), Elsevier Science B.V., New York, 365-372.

[99] Shimizu H, Kozaki Y, kodama H, Shioya H. 1998. Maximum production strategy for biodegradable copolymer PHBV in fed-batch culture of Alcaligenesis eutrophus, Biotechnol. Bioeng., 62(5):518-525.

[100] Hossein Hariri Asli (2023) Reclaimed Asphalt Pavement (RAP) Based on Geospatial Information System (GIS) and Networked Sensors Modelling, New Advances in Materials Technologies; Experimental Characterizations, Theoretical Modeling, and Field Practices, Hard ISBN: 9781774914847, https://www.appleacademicpress.com/new-advances-in-material-technologies-experimental-characterizations-theoretical-modeling-and-field-practices/9781774914847

[101] Hossein Hariri Asli, (2023) Applications of Networked Sensors and the Internet of Things (IoT) for Wastewater Treatment, Sustainable Water Treatment and Ecosystem Protection Strategies, Hard ISBN: 9781774915189, https://www.appleacademicpress.com/sustainable-water-treatment-and-ecosystem-protection-strategies-/9781774915189

[102] Steinbuchel, A. (1996). PHB and other polyhydroxyalkanoic acids. In: Biotechnology, Rehm, H.J. and Reed, G. (eds.) VCH, New York, 403-464.

[103] Hariri Asli, K., Soltan, A. O., Thomas, S., Deepu, A., Hariri Asli, H. (2017). Handbook of Research for Fluid & Solid Mechanics, Published by the Apple Academic Press, Inc., ISBN 9781771885010, USA, Canada, https://doi.org/10.1201/9781315365701 https://www.taylorfrancis.com/books/edit/10.1201/9781315365701/handbook-research-fluid-solid-mechanics-kaveh-hariri-asli-soltan-ali-ogli-aliyev-sabu-thomas-deepu-gopakumar

[104] Hossein Hariri Asli Ali Pourhashemi, Ann Rose Abraham, A. K. Haghi (2023) New Advances in Materials Technologies; Experimental Characterizations, Theoretical Modeling, and Field Practices, Hard ISBN: 9781774914847, https://www.appleacademicpress.com/new-advances-in-material-technologies-experimental-characterizations-theoretical-modeling-and-field-practices/9781774914847

References

[105] Suzuki T, Yamane T, Shimizu S. 1989. Mass production of PHB by fully automatic fed-batch culture of methylotroph. Appl. Microbiol. Biotechnol, 23:322-329.

[106] Takabatake H., Satoh H., Mino T. and Matsuo. (2000), Recovery of biodegradable plastics from activated sludge process, Wat. Sci. Tech, 42(3-4), 351-356.

[107] Tsuchikura, K. (1994). BIOPOL Properties and processing, in: Biodegradable plastics and polymers, Doi, Y. and Fukuda, K. (Eds.). Elsevier Science B.V., New York, 362-364.

[108] Van Groenestijn, J. W., Deinema, M.H., and Zehnder, A.J.B. (1987). ATP Production from polyphosphate in Acinetobacter strain 210A. Arch Microbial. 148, 14-19.

[109] Hossein Hariri Asli, (2023) Applications of Networked Sensors and the Internet of Things (IoT) for Water Treatment, Sustainable Water Treatment and Ecosystem Protection Strategies, Hard ISBN: 9781774915189, https://www.appleacademicpress.com/sustainable-water-treatment-and-ecosystem-protection-strategies-/9781774915189

[110] Hossein Hariri Asli, (2023) Modeling of Corrosion for Water System by Networked Sensors and the Internet of Things (IoT) in Compliance with Geography Information System (GIS), Sustainable Water Treatment and Ecosystem Protection Strategies, Hard ISBN: 9781774915189, https://www.appleacademicpress.com/sustainable-water-treatment-and-ecosystem-protection-strategies-/9781774915189

[111] Yamane, T. (1993). Yield of Poly-D (-)-3hydroxybutyrate from various carbon sources: A Theoretical Study. Biotechnology and Bioengineering. 41(1), 165-170.

[112] Asli, H. H. and Arabani, M. 2022 Analysis of Strain and Failure of Asphalt Pavement. Computational Research Progress in Applied Science & Engineering. Transactions of Civil and Environmental Engineering 8. 1–11. Article ID: 2250. https://doi.org/10.52547/crpase.8.1.2250.

[113] Kian Hariri Asli; Kaveh Hariri Asli (2023) Minimum night flow (MNF) and corrosion control in compliance with internet of things (IoT) for water systems, Water Practice and Technology, wpt2023012, https://doi.org/10.2166/wpt.2023.012, https://iwaponline.com/wpt/article/doi/10.2166/wpt.2023.012/93513/Minimum-night-flow-MNF-and-corrosion-control-in

[114] Asano T, Water from Wastewater, The Dependable Water Resource (The 2001 Stockholm Water Prize Laureate Lecture). Water Science and Technology, 2002, 45(1), 23-33.

[115] Asli KiH, Asli KaH. Smart Water System and Internet of Things. J Mod Ind Manuf, 2023; 2: 5. DOI: 10.53964/jmim.2023005, DOI: 10.53964/jmim.2023005.

References

[116] Tapia EM, Intille SS, Larson K, "Portable wireless sensors for object usage sensing in the home: Challenges and practicalities," in Proceedings of the European Ambient Intelligence Conference. vol. LNCS 4794 Berlin Heidelberg: Springer-Verlag 2007

[117] Amritanshu Shukla, Kian Hariri Asli, Neha Kanwar Rawat, Ann Rose Abraham, A. K. Haghi, Technological Advancement in Clean Energy Production ,(Apple Academic Press, Hard ISBN: 9781774915585), (2024), https://www.appleacademicpress.com/title.php?id=1416

[118] Ali Pourhashemi, Kian Hariri Asli, Ann Rose Abraham, A. K. Haghi, Sustainable Water Engineering; Smart and Emerging Technologies, (Apple Academic Press, Hard ISBN: 9781774915714), (2024), https://appleacademicpress.com/sustainable-water-engineering-smart-and-emerging-technologies/9781774915714

[119] Ann Rose Abraham, Heru Susanto, A. K. Haghi, Kian Hariri Asli, Sustainability in Energy and Environment; Engineered Materials and Smart Computational Techniques",(Apple Academic Press, Hard ISBN: 9781774916209), (2024), https://www.appleacademicpress.com/sustainability-in-energy-and-environment-engineered-materials-and-smart-computational-techniques/9781774916209#bios

[120] Sonia Khanna, Ann Rose Abraham, Kian Hariri Asli, A. K. Haghi, Sustainable Environmental Engineering; Water Security, Energy Conservation, and Green Processes", (Apple Academic Press, Hard ISBN: 9781774916902), (2024), https://www.appleacademicpress.com/sustainable-environmental-engineering-water-security-energy-conservation-and-green-processes/9781774916902

[121] Roman R, ALcaraz C, Lopez J, , et al., "Key Management Systems for Sensor Networks in the Context of the Internet of Things," Computers & Electrical Engineering, 2011,37(2), 147-159

[122] Offermans A, Haasnoot M, Valkering P, A method to explore social response for sustainable water management strategies under changing conditions, Sustainable Development, 2011,19(5), 312-324.

[123] Ann Rose Abraham, Heru Susanto AK, Haghi, et al., Sustainability in Energy and Environment Engineered Materials and Smart Computational Techniques, Apple Academic Press,2024, Hard ISBN: 9781774916209, https://www.appleacademicpress.com/sustainability-in-energy-and-environment-engineered-materials-and-smart-computational-techniques/9781774916209#bios

[124] Shahhosseini, S., "A fed-batch model for PHA production using Alcaligenes eutrophus", M.S. Thesis, Department of Chemical Engineering, The University of Queensland, Australia, 1994.

[125] Haselbach, L., Adesina, M., Muppavarapu, N. *et al.* Spatially estimating flooding depths from damage reports. *Nat Hazards* **117**, 1633–1645 (2023). https://doi.org/10.1007/s11069-023-05921-2.

References

[126]   Weidner, Jeffrey; Collins, Jin; Benitez, Mariana; Adesina, Mubarak; Lozoya, Christian; 2019, Development of a Robust Framework for Assessing Bridge Performance using a Multiple Model Approach, University of Texas at El Paso. Department of Civil Engineering, Report Number: CAIT-UTC-NC39, https://rosap.ntl.bts.gov/view/dot/48948.

Index

A
Activated sludge 11, 15, 18, 33, 40, 42
Aerobic 9, 11, 19, 20, 28, 35
Aerobic reactor 9, 20, 29
Anaerobic tank 9, 10, 20, 29

B
Biochemical oxygen demand (BOD) 11
Bio-plastic 31

C
Closed loop control system (CLCS) 28
Correlation 20, 27, 31

D
Data intercommunication 29
Degradable plastics 31, 34, 37, 39, 40
Dependent variable 20, 24, 25, 26, 31
Detention time 12, 15, 19

F
Flow rate 13

G
Geographic Information System (GIS) 40
Greywater 31

I
Independent variable 20, 25, 26, 31
Influent 11, 12, 13, 33
Internet of Things (IoT) 9, 20, 28, 29, 39, 41, 43

M
Micro-organism 11, 12, 31

N
Networked sensors 20, 39, 40, 41

O
Oxidation reduction potential (ORP) 28

P
Pollution detention 12
Polyhydroxyalkanoates (PHAs) 8, 9, 31, 34, 35, 38
Porous diffusers 12
Pure O2 12, 15

R
Regression analysis 20, 21, 23, 31
Remote sensing (RS) 9, 20, 28, 29

S
Scatter diagram 19, 20, 21, 22, 23, 31
Sedimentation tank 9, 17, 20, 29

W
Wastewater reuse 7, 8, 9, 15, 17, 19, 26, 28, 31
Wastewater treatment 3, 40, 41
Water loss 2, 5, 7